Investigations and Problem Solving

Olwen El-Naggar

Published in 1995

© Olwen El-Naggar

All rights reserved. Apart from the Resources Section no other part of this publication may be reproduced, stored in a retrieval system, or transmitted in any form or by any means, electronic, mechanical, photocopying, recording or otherwise, without the prior written permission of the publisher.

The right of the Author to be identified as Author of this work has been asserted by her in accordance with the Copyright, Designs and Patents Act 1988.

Published by QEd, The Rom Building, Eastern Avenue, Lichfield, Staffs. WS13 6RN

ISBN 1 898873 02 X

Further copies of this book may be obtained from:
QEd, The Rom Building, Eastern Avenue, Lichfield, Staffordshire WS13 6RN

Cover design by Graphic Images.
Typeset in Century Schoolbook. Printed in the United Kingdom by J.H. Brookes.

Contents

Activity	Programme of Study	Page
Introduction		4
Addition and Subtraction	Number	7, 9
Show Me Sums	Number	11
Looking for Patterns	Shape, Space and Measures	13
Addition Towers	Number	15
Comparison [1]	Shape, Space and Measures	17
Comparison [2]	Shape, Space and Measures	19
Jumps	Shape, Space and Measures	21
Shapes [1]	Shape, Space and Measures	23
Shapes [2]	Shape, Space and Measures	25
Pendulums	Shape, Space and Measures	27
Magic Domino Squares	Number	29
Making Shapes	Shape, Space and Measures	31
Footprints	Shape, Space and Measures	33
Strides	Shape, Space and Measures	35
Difference Towers	Number	37
Symbol Towers	Number	39
Addition Squares [1]	Number	41
Addition Squares [2]	Number	43
Playing Cards	Number	45
Halving [1]	Number, Shape, Space and Measures	47
Halving [2]	Number, Shape, Space and Measures	49
Names and Numbers	Number	51
Length	Shape, Space and Measures	53
Making Numbers	Number	55
Grid Sums	Number	57
21 Out	Number	59
Take Six Dice	Number	61
Personal Data File	Number, Shape, Space and Measures	63
Down On The Farm	Number	65
Take Six Dots	Shape, Space and Measures	67
Telephone Numbers [1]	Number	69
Telephone Numbers [2]	Number	71
RESOURCES		73 - 80

Introduction

'Teachers devote much time to work with numbers and the practice of four rules and many children achieve a satisfactory level of competence in this field, but few have sufficient opportunity for learning how to apply the skills they acquire to the solving of problems .....Too few schools make opportunities for the development and extension of mathematical understanding which arises in children's play, in their interest and in work in other parts of the curriculum.' (Cockroft Report: *Mathematics Counts* para 4:9, 1983)

The following collection of activities has been developed through trialling the material in schools. They are written for exposition by the teacher and, where necessary, pupil worksheets are provided.

For many pupils, including those with special needs, exposition by the teacher means more than just reading the questions to pupils with poor literacy skills, or explaining the computation involved to pupils who have not mastered concrete/abstract transference. It means offering the activity at a level commensurate to the pupils' level of development in abstract thinking. Many examples of mathematics given in text books do not address this development and pupils with learning difficulties are often excluded from problem solving.

This book offers guidance on differentiating activities to make them accessible to pupils who need a staged approach to problem solving. The four stages used are an adaptation of Zoltan Dienes' six stages (Z.P. Dienes: *The Six Stages In The Process Of Learning Mathematics* 1973 NFER).

Stage 1: Physical Experience

Pupils need to be placed in an environment through which they will learn the mathematics involved. They need to be given experience of practical examples to realise links of abstraction. Experience with a wide variety of materials using a multisensory approach facilitates the development of strong concepts.

Stage 2: Communication

It is through communicating ideas that pupils develop logical thinking. Mathematics language is more than a list of appropriate words; it is the ability to present mathematical ideas in logical sequence. At this stage pupils talk about the experiences of Stage 1.

Stage 3: Representation

Before abstraction pupils need a method of representation, either concrete or pictorial including graphical representation. These images will be recalled when solving future mathematical problems.

Stage 4: Symbolisation

Symbols are a shorthand way of representing all previous experience. It is important to remember that they are not the concept, they are representative of it. Without experiencing the previous three stages most pupils encounter great difficulty with symbols.

Where appropriate suggestions for differentiating activities through this four stage process is offered in the accompanying Teachers' Notes.

References to National Curriculum Programmes of Study and any essential material are listed at the beginning of each page.

The order in which each activity is approached should be dictated by the immediate needs of pupils. Select activities to match those needs and supplement work in your school scheme. Although pupils may evidence a particular level of achievement, revisiting previous levels, to strengthen concepts and build confidence, is a necessary part of mathematics development.

Organisation differs from school to school, possibly from class to class. Use the activities in the way best suited to your needs, remembering that:

1. Mathematics is an enjoyable subject.
2. It is the process pupils go through which is important and any valid solution should be accepted.

All the activities in this book offer opportunities to make and monitor decisions to solve problems, develop mathematical communications, and develop mathematical reasoning.

Teachers' Notes

Addition and Subtraction

This activity is presented at both the physical representation and communication stage. It may be extended to cover representation and symbolisation.

Representation

- Pupils may draw what they did.

Symbolisation

- Pupils may use number symbols to represent what they did.

    For example, 6 + 1 = 7

Number

Addition and Subtraction

Skill Developing an ability to solve numerical problems (addition and subtraction to 10).

Materials Chalk board, a bag, 10 objects e.g. beads or buttons, pencil and paper.

Activity

An activity for 3 pupils.

Pupil number 1 writes a number less than ten on the board.

Pupil number 2 takes a handful of objects from the bag (without looking at them) and places them in front of Pupil number 3.

Pupil number 3 adds or subtracts the correct number of objects to make the number written on the board.

Pupils are then asked to make a number story saying what they did.

Teachers' Notes

Addition and Subtraction

This activity offers opportunity at the representation stage and may be preceded by:

Physical experience

- Give each pupil 20 multilink cubes (or building bricks).
- Make some gummed labels 1 to 20 and invite pupils to stick one number on each cube or brick.
- Practice ordering the bricks or blocks (ordinal number).

Communication

- Talk about ordering numbers in terms of which number is first/last/between/next to/before/after.

and extended to:

Symbolisation

- Number stories can be represented by arrows on a number line.

Number

Addition and Subtraction

Skill Developing an ability to understand place value (counting).

Materials A number line (houses resources sheet - page 78).

Activity

Give each pupil a number line, in the form of a row of houses. Now ask him or her to follow this story on the line.

My street is a very unusual street as all the houses are on one side. On the other side of the street is a park. The street starts with house number 1.

Put your finger on number 1.

Next door to number 1 is number 2, and then 3, and 4, and 5, and 6. I live in house number 3.

Show me house number 3.

My gran lives two houses higher up the street than me.

Show me where she lives.

Four houses higher up the street is where the little brown dog lives. His name is Fido.

What is the number on the door of his house?

Teachers' Notes

Show Me Sums

This activity is presented at the representation stage.

Communication and symbolisation are not appropriate to this activity; it provides opportunity for pupils to:

(a) listen to what they are being asked to do;

(b) decide what to do;

(c) do it.

The omission of both spoken and written answers encourages all children to participate.

You may need to make some ground rules, for example:

- a single star displayed on the chalkboard means that only the teacher may speak;
- the questions are only given once, everybody has to concentrate on listening;
- nobody must shout the answers out. All that they should do is show the answer.

This is a large group/whole class activity enabling every pupil to respond to the question at his or her own speed.

Number

Show Me Sums

Skills To develop an ability to understand problems at the concrete stage;
Develop problem solving strategies ... decide, do and check.

Materials Twenty multilink cubes and a piece of paper for each child.

Activity

This activity allows for individual pace, giving each pupil time to realise an answer.

The following statements should be read to a whole class/large group. After each one allow time for all children to show their answers.

- David put seven glasses of orange juice on the table. **Show me.**
- Aziza had six friends. **Show me.**
- The cat had eight kittens. **Show me.**
- Five children went to the circus. **Show me.**
- Nine cars drove into the car park. **Show me.**
- Simon put seven cakes on a plate. **Show me.**
- The dog had six puppies, three were brown and three were black. **Show me.**
- Six ice lollies in a row. The first and the fourth melted. **Show me.**
- Laura put six boxes on the table, she put them in three equal piles. **Show me.**
- Tom baked three cakes, each cake had two layers. **Show me.**
- Eight birds on a fence. The third and the fourth flew away. **Show me.**
- Six houses on one side of the road and four houses on the other side of the road. **Show me.**

Teachers' Notes

Looking For Patterns

The idea behind the resource sheet is to check that pupils understand that patterns can be continued from either end.

Shape, Space and Measures

Looking for Patterns

Skills Developing an ability to compare and order objects without measuring, using appropriate language;
Recognise simple patterns and relationships and make predictions about them.

Materials Two different types of plastic containers (four of each), 4 pieces of 10 x 15cm card, a felt pen or the **resources sheet** (page 79).

Activity

Place the containers in order to form a pattern.

e.g.

- Ask pupils to continue the pattern and to make as many different patterns as they can.
- Talk about these patterns.
- Make pictorial representations of the containers on the card.
- Place three of the container cards in order.
- Ask pupils to show where the fourth card should go.

Teachers' Notes

Addition Towers

This activity is presented at both the representation and symbolic stage. Communication between pupils is encouraged through group work.

Number

Addition Towers

Skill   Developing an ability to explore and record number patterns.

Materials   Unifix or multilink cubes, paper and pencils.

Activity

This activity can be presented in several ways offering learning situations at different levels.

- Place 20 cubes on the table and invite pupils to make 5 towers. Each tower should be of a different colour. The towers do not necessarily have to be of the same height.
- Pupils now combine the stacks attempting to make each of the numbers 1 to 10.
- Each time a pupil makes a number he or she records it.

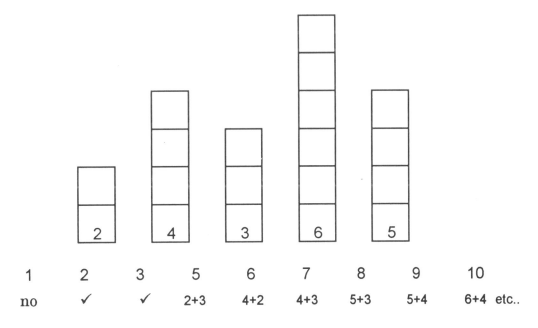

| 1 | 2 | 3 | 5 | 6 | 7 | 8 | 9 | 10 |
| no | ✓ | ✓ | 2+3 | 4+2 | 4+3 | 5+3 | 5+4 | 6+4 etc.. |

This activity should be repeated many times using different size towers.

Teachers' Notes

Comparison (1)

This collection of activities offers opportunity in physical experience, language and pictorial representation.

There is no symbolic recording at this stage.

Shape, Space and Measures

Comparison [1]

Skills   Developing an ability to compare objects using appropriate language;
Use non standard units of measurement.

Materials   A bucket, a cup, a box, cardboard, a felt tipped pen.

Activity

1. Make some large labels TALL/SHORT.

- Have pupils label pairs of objects around the room as **TALL/SHORT**.

2. Find a box.

- Collect objects that hold less than the box.
- How do you know that they hold less than the box?

3. Find a cup.

- Which holds more water? ... A bucket or a cup? How do you know?
- Collect things that hold less than the cup.
- How do you know that they hold less than the cup?

4. Make wall charts for the following.

- Objects that are taller than a spider.
- Objects that are taller than me.
- Objects that are shorter than me.
- Objects that are taller than a house.

Teachers' Notes

Comparison (2)

This collection of activities offers opportunity in physical experience, language and pictorial representation.

There is no symbolic recording at this stage.

Shape, Space and Measures

Comparison [2]

Skills Developing an ability to compare objects using appropriate language;
Use non standard units of length.

Materials A small number of multilink cubes, a pencil, a hoop, a straw, a ball of Plasticine, paper, scissors.

Activity

1. Using multilink cubes.

- Build a tower as tall as your pencil.
- How many cubes did it take?
- Draw your pencil and the tower.

2. Using a hoop and a straw.

- Find five things **shorter** than the straw and put them inside the hoop.
- Draw them all in your book.

3. Using some Plasticine.

- Make a snake.
- Make a **longer** snake.
- Draw your snakes in your book and give them each a name.
- What is the name of the **longest** snake?
- Which is your favourite snake? Why?

4. Shoes and feet.

- Find five things shorter than your shoe and draw them in your book.
- Have four or five pupils trace one of his or her feet onto a piece of paper and cut it out.
- Place them on the floor to form a set of feet.
- Sort them into two sets, large feet and small feet.
- Now place them in order, **smallest first.**
- Now place them in order, **largest first.**

Teachers' Notes

> ## Jumps
>
> This collection of activities offers opportunity in physical experience, language, and pictorial representation in the form of data handling activities.
>
> There is no symbolic recording at this stage.

Shape, Space and Measures

Jumps

Skills Compare and order without measuring, and use appropriate language;
Use language to describe position.

Materials Three people, a large piece of paper, felt tips, pencils, room to jump.

Activity

1. The Jumps

- Select three people and a place to jump.
- Mark the starting point.
- Ask each of the people to jump in turn.
- Now fill in the chart.

The *longest* jump was	
The *shortest* jump was	
Who came *first*?	
Who came *second*?	
Who came *third*?	

2. Backward Jumps.

- Do the same for backward jumps.
- What did you notice?

3. Other Jumps.

- Ask the children to record any other jumps they can think of.
- Which jumps were the longest?
- Which were the hardest jumps to do?

Teachers' Notes

Shapes (1)

This activity offers opportunity for physical experience, communication and pictorial representation.

Symbols used for shape at this stage are alphabetic in the form of shape names.

Shape, Space and Measures

Shapes [1]

Skill Developing an ability to sort and classify 3D shapes and describe their properties.

Materials A box of solid shapes, a large sheet of paper, pencils, a board, or sloping surface (for rolling shapes on) Resource page 80.

Activity

1. Sorting and classifying shapes.

- Collect the following shapes from the box:

    a sphere

    a cylinder

    a cone

    a cube

    a cuboid (rectangular prism)

- Which of these shapes is best for rolling?

    Why?

- Which is best for building?

    Why?

- Which shape can only roll in a circle?

2. Once the above activities have been completed fill in the information on the chart (Resources sheet page 80).

Teachers' Notes

Shapes (2)

Tip: If you store your multilink cubes close to a radiator they will remain pliable.

By using the same ball of Plasticine pupils can feel the physical changes as they make new shapes:

- compressing the sphere to make the six faces of a cube;
- stretching the cube to make a cuboid;
- the cone is a challenge but is possible.

Shape, Space and Measures

Shapes [2]

Skill Developing an ability to sort and classify 3D shapes and describe their properties.

Materials Multilink cubes, a ball of Plasticine, a cloth bag.

Activity

1. Using multilink cubes.

- Take four cubes and join them together anyway you like.
- How many different ways can you join four cubes. Are you sure they are all different even if you turn them round?
- Now try buildings with five cubes.

2. Using Plasticine.

- Take a ball of Plasticine and make a sphere. How do you know it is a sphere?
- Use that same ball of Plasticine to make a cube. How do you know it is a cube?
- Use the same ball of Plasticine to make a cuboid. How do you know it is a cuboid?
- Can you make a cone?

3. Using a bag.

- Take a small bag made of non transparent material and allow pupils to examine it inside and out.
- Place a solid shape in the bag (secretly).
- Invite a pupil to feel the shape in the bag and describe what they feel to the rest of the group.
- A variation of this activity is to place a number of solids in the bag and invite each pupil to feel the shapes and name as many as they can. The winner is the person who correctly names the most.

Teachers' Notes

Pendulums

This activity offers opportunity in physical experience and communication it can be extended to pictorial representation of the experiments. These could be accompanied by written explanations.

Shape, Space and Measures

Pendulums

Skill   Developing an ability to compare speed and length, using appropriate language;
Use mathematical language;
Use mathematical reasoning.

Materials   String, a ball of Plasticine, scissors, tape.

Activity

- Fix a small ball of Plasticine firmly to one end of a piece of string.
- Tape the string underneath a table, making sure the Plasticine does not touch the floor.
- Swing the pendulum gently and count the swings whilst a friend writes his or her first name.

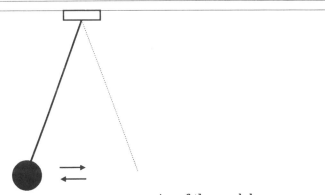

one swing of the pendulum

- Make the Plasticine blob larger. Count the swings whilst the same friend writes his/her first name.
- What will happen if you make the blob smaller?
- What will happen if you make the string shorter? Check your answer.
- Make some more pendulums. What could you use instead of Plasticine?

Teachers' Notes

Magic Domino Squares

This activity offers opportunity at the representation stage and can be preceded by:

Physical experience:

Place a set of dominos on the table and encourage free play (about 5 minutes). This gives pupils time to become familiar with the numbering system on the dominos.

Communication:

Draw the group together and invite discussion.

Questions to ask:

- What did you notice about the dominos?
- Are they all the same shape?
- Are they all the same size?
- Do they all have spots on them?
- What is the highest number of spots on any single domino?
- What is the lowest number of spots?

and extended to:

Symbolisation:

Solutions can be recorded in symbolic terms:

$5 + 1 + 2 = 8$

$2 + 6 + 0 = 8$

$3 + 5 + 0 = 8$

$5 + 0 + 3 = 8$

Number

Magic Domino Squares

Skill Developing an ability to solve numerical problems in addition.

Material A set of dominos.

Activity

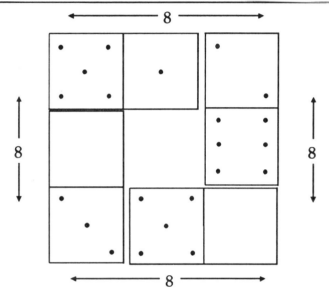

This is a magic domino square made up of four ordinary dominoes, arranged so that the total number of dots on each side of the square are the same.

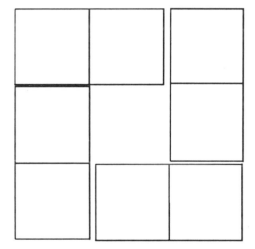

- Using ordinary dominoes with numbers 0 to 6 on them, complete this square for a magic number of 9.
- Now make a magic domino square for 10.
- Make some magic domino squares of your own.

Teachers' Notes

> # Making Shapes
>
> These activities offer opportunity in physical experience, language and pictorial representation.
>
> There is no symbolic recording for this activity.

Shape, Space and Measures

Making Shapes

Skill Developing an ability to use mathematical language to describe shapes.

Materials For 3D shapes you will need junk materials, paints, glue, scissors.
For 2D shapes you will need paper, pencils, paints or felt tips.

Activity This activity is designed to draw the attention of pupils to the fact that not all shapes have names. Shapes can be either 3D or 2D.

- Make a shape which you think would go very fast.
  Draw it in your book and write *fast* on it.

- Make a shape which you think would move very slowly.
  Draw it in your book and write *slow* on it.

- Make a shape which would be very safe.
  Draw it in your book and write *safe* on it.

- Make a shape which would be unsafe.
  Draw it in your book and write *unsafe* on it.

- Make a shape which you think is very ugly. Draw it in your book and write *ugly* on it.

- Make a shape which you think is very pretty.
  Draw it in your book and write *pretty* on it.

- You think of some more ....

Teachers' Notes

Footprints

These activities offer opportunity in physical experience, language and graphical representation (data handling work may be mounted on display boards).

There is no symbolic recording for this activity.

Shape, Space and Measures

Footprints

Skills Developing an ability to use non-standard measures in length and area; and compare size using appropriate language.

Materials Scissors, newspaper.

Activities

- Draw around your foot and cut out your footprint. Find a friend with the same size footprint.
  Now use your two footprints to find how wide the classroom is.

- Find six things in your classroom that are longer than your footprint. Now draw them.

- How many of your footprints would it take to cover the top of your table?

- Draw the foot prints of your family and put them in order, largest first.

- Do footprints make a good measure?

- Draw around both your feet and cut the shapes out.
  Are both your feet the same size?
  How many people in your group have both feet the same size?

Teachers' Notes

> # Strides
>
> These activities offer opportunity in physical experience, language and symbolic representation.
>
> For pupils working at the representation stage, cubes, counters or bottle tops could be used (one to represent each stride) and comparisons drawn. These could be recorded on comparison charts.

Shape, Space and Measures

Strides

Skill Developing an ability to estimate;
Use non-standard measures in length;
Compare length.

Materials Pencil and paper or a prepared chart.

Activity

Estimate *first* and *then* measure

- How many strides will it take to cross the classroom?
- Will your teacher take fewer strides than you? Why?
- How many strides to go all around the classroom?

Make a table to record your results. It could look like this:

How many strides?	estimate	measure
to cross the room		
around the room		

- Use your stride to measure something else and put it on your chart.
- Use a piece of string, or some multilink cubes to measure your stride.
- How many of your friends have a shorter stride than you?

Teachers' Notes

> # Difference Towers
>
> This activity offers opportunity for physical experience, communication, concrete representation and symbolic representation.

Number

Difference Towers

Skill   Developing an ability to compare two numbers to find the difference.

Materials   Unifix or multilink cubes, pencils and paper.

Activities

- Place 20 cubes on the table and invite pupils to make 5 towers.
  The towers do not necessarily need to be made of the same colour cubes.
  The towers do not necessarily need to be of the same height.

- Place any two towers next to each other and discuss the difference in terms of more or less.

- Record the results.

The difference between 5 and 2 is 3

2 is 3 less than 5   5 is 3 more than 2

What is the difference between 3 and 4?

What is the difference between 2 and 5?

Teachers' Notes

Symbol Towers

This activity offers opportunity in connecting concrete experience to symbolic recording.

Number

Symbol Towers

Skill Developing an ability to understand the use of a symbol to stand for an unknown number.

Materials 20 unifix or multilink cubes, pencils and paper.

Activities

- Place the 20 cubes on the table and invite the pupils to build 5 towers, one each of the following heights: 2 cubes, 3 cubes, 4 cubes, 5 cubes, 6 cubes. (The towers do not necessarily have to be of the same colour).

- Use the cubes to find the correct numbers for the boxes.

$$3 + \square = 5 \qquad 4 + \square = 6$$

$$2 + \square = 5 \qquad 2 + \square = 6$$

- Increase the number of towers to make all the numbers to 10. Now try these.

$$1 + \square = 7 \qquad 8 + \square = 9$$

$$5 + \square = 8 \qquad 7 + \square = 9$$

$$6 + \square = 10 \qquad 6 + \square = 9$$

- Make some symbol sums of your own.

Teachers' Notes

Addition Squares (1)

This activity is at the symbolic stage.

For pupils who need a staged approach:

- Start with a 2 x 2 square and assist them to fill it in.
- When they have mastered this, progress to a 5 x 5 square.
- It is sometimes helpful to pupils if the position of the 2 x 2 square is highlighted in the 5 x 5 square.
- Some of the numbers could already be filled in on the 10 x 10 square.

Number

Addition Squares [1]

Skills   Developing an ability to solve number problems in addition; Explore patterns in addition and subtraction.

Material   Grid squares.

Activities

- Fill in the missing numbers.

2		
1		
+	1	2

- Fill in the missing numbers.
- Is there a pattern?
- Where does the row of 6s start? What about the 5s and 4s and 3s?
- What do you notice about the starting and finishing numbers?

5	6				
4					
3			6		
2				6	
1					
+	1	2	3	4	5

- Fill in the missing numbers.
- Find the patterns.
- Use your addition table to find the answer to these sums:

    $7 + 7 = ?$
    $3 + 8 = ?$
    $5 + 5 = ?$

Make up 10 more addition sums.

10									20
9									
8									
7									
6									
5	6				10				
4		6							
3			6				10		
2				6				10	
1					6				
+	1	2	3	4	5	6	7	8	9	10

Page 41

Number

Addition Squares [2]

Activities

- Use your addition square to find the missing numbers.

    $8 + \square = 11$ $8 + \square = 10$
    $5 + \square = 3$ $5 + \square = 7$
    $3 + \square = 9$ $3 + \square = 8$

- Make 4 more missing number sums and write them in your book.

Try this:

- Use your addition square and add any two numbers together.
  For example $5 + 3 = 8$.

- Put your finger on any of the 5's in the square.
  Now move it three spaces; each move may be either to the right or upwards.

4	5	6	7	8	9
3	4	5	6	7	8
2	3	4	5	6	7
1	2	3	4	5	6

- What do you notice?
  Try this again, starting with another 5.
  Does it always work?

- Make another addition sum and try again.

- Does it work with any addition sum?

- How could you make it work for subtraction sums?

Teachers' Notes

Playing Cards

This activity offers opportunity for physical experience, communication and data handling. Pupils may wish to record their results.

Number

Playing Cards

Skills Developing an ability to choose criteria for sorting a set of objects, record results;
Design a data collection sheet;
Construct and interpret frequency tables and block graphs;
Observe odd and even numbers.

Materials A pack of playing cards, pencil, felt tips, paper (some squared).

Activities

In a pack of playing cards no two of them are alike. They belong to four different suites.

- How many cards are there in each suite?
- How many cards in a pack show odd numbers?
- How many cards in a pack show even numbers?
- How many cards are court cards, that is, show pictures of a Jack, Queen or King?
- Make a chart to show all this information.

Use your chart to find the following:

- Are there more odd than even cards?
- How many more even cards than court cards are there?
- Are there more women than men on the court cards?

Add some information of your own to the chart.

- Use the pack of cards to find out how many ways you can make 21.
  An Ace can be either 1 or 11.
  A Court card is 10.

Teachers' Notes

Halving (1)

This activity offers opportunity for physical experience, communication and pictorial representation. Symbolic representation is not necessary at this stage.

Number, Shape, Space and Measures

Halving [1]

Skill   Developing an ability to understand the meaning of *a half*.

Materials   Squared paper (resources sheet page 77), felt pens.

Activities

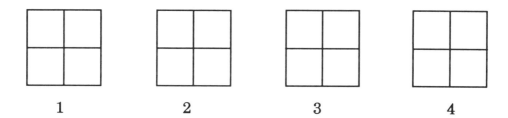

Here are some squares with 4 smaller squares inside them.

- Use one colour only and colour half of square number 1.
- Colour half of square number 2 in a different way.
- Try to colour half of each square but do each one in a different way.

Make some different size squares and colour half.

- How can you divide your class in half?
- If there is an odd number of children what will happen to the one left over?

Look at the rectangles

- Is half of them black?
- How can you find out?

Page 47

Teachers' Notes

Halving (2)

These activities offer opportunity for physical experience, communication, representation and symbolisation.

Number, Shape, Space and Measures

Halving [2]

Materials 2 plastic glasses, a small bag of sand, scales, a length of string, a ball of Plasticine;
Paper and pencils, scissors.

Activities

- How would you divide a glass of water into halves?
- How would you halve a bag of sand?
- Find half a length of string.
  Do you have to measure?
- Find half a ball of Plasticine.
- Collect together all the things you can find half of?

Cutting in half

- Take a square of paper. Cut the square in half.
  How many pieces have you got?
- Cut each piece in half.
  How many pieces have you got now?
- Cut each piece in half again. Keep doing this and record your results each time.

	number of pieces
1st cut	2
2nd cut	4
3rd cut	8
4th cut	?

Can you see a pattern?

- What will the next number be?
- And the next?

Teachers' Notes

Names and Numbers

This activity is presented at the representation stage and progresses to symbolisation. It could be preceded by:

Physical Experience

- Inviting pupils to place coloured counters on the resource sheets to make patterns, using two colours then three, four and five colours.

Communication

Questions to ask:

- What happened when you used three colours?
- What else did you notice?
- Which is your favourite pattern? Why?

Number

Names and Numbers

Skill Developing an ability to recognise patterns and relationships and make predictions about them.

Materials A set of squares 2 x 2, 3 x 3, 4 x 4, 5 x 5, 6 x 6, 7 x 7, per child (resources sheet page 75).

Activities

- Write your name continuously in the squares, beginning each new set of squares with the first letter of your name. Do not leave any spaces.

J	o
h	n

J	o	h
n	J	o
h	n	J

L	a	u	r
a	L	a	u
r	a	L	a
u	r	a	L

- Now colour the first letter of your name.
- Can you see a pattern?
- Have any of your friends got the same pattern?
- Do you know why?
- What size square would you need to make the same patterns with these names: **Ahmed, Mary, Rachel**?
- Write down the names of 2 members of your family. What size squares do you need to make patterns with their names?
- Now fill some squares using numbers 1 to 6.
- Can you see a pattern?
- Try some different numbers and look for patterns.

Page 51

Teachers' Notes

> # Length
>
> These activities offer opportunity for physical experience, communication and representation. Symbolisation is not necessary at this stage.

Number, Shape, Space and Measures

Length

Skill Developing an ability to choose and use simple measuring instruments, read and interpret scales with appropriate accuracy.

Materials Rulers, tape measures, height charts, string, foot gauge, callipers, paper and pencils.

Activities

Personal Data

- Make a measurement chart to show the following personal data.

part of body	cm
neck	
waist	
foot	
nose	
height	
ear	

How do we measure a curve?

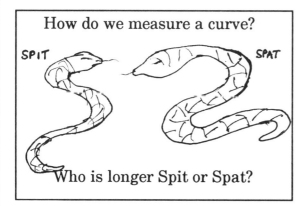

Who is longer Spit or Spat?

Which of these lines is the longest?
Estimate and measure.
What did you find?

Half of this line is hidden.

———————

How long is the whole line?

Teachers' Notes

Making Numbers

This activity is presented at the symbolic stage and may be preceded by:

Physical experience

- Ask pupils to select two objects e.g. two different coloured counters/buttons/cubes.

- Ask pupils how many ways they can arrange their two objects.

- Now ask them to take three objects and see how many different ways they can arrange these.

- Pupils could be encouraged to investigate further numbers of objects.

Communication

- Talk about the results.

Representation

- Pupils should be encouraged to record their results in any way they wish. The results should then be discussed.

Number

Making Numbers

Skills Developing an ability to understand that the position of a digit signifies its value;
Explore number patterns;
Observe odd and even numbers.

Materials Cards, pencils.

Activities

Make a set of cards with these numbers on them:

| 1 | | 2 | | 3 |

- Choose two cards to make a number. For example you could make:

| 2 | 3 |

- How many different two card numbers can you make.
  Write them all down.
- Put a circle around the even numbers.

Make another set of cards with these numbers on them:

| 7 | | 5 | | 9 |

- How many numbers can you make this time?
- Put a circle around the odd numbers.
- Explain what happens. Write about it in your book.
- Can you choose 3 cards so that your set of numbers will all be even? **Yes** or **No**.
- Make **one** more card so that you can.
- Can you tell how many before you write them down?

Teachers' Notes

Grid Sums

This activity is at the symbolic stage and should be given to pupils who are comfortable with this stage.

It should be preceded by discussion about rows and columns.

Number

Grid Sums

Skills   Developing an ability to explain and record findings;
Understand place value.

Materials   A die, (resources sheet 76), pencils, a calculator (optional).

Activities

Use the 4 by 4 squares (Resource sheet on page 76)

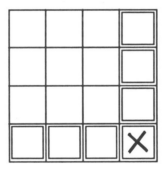

- Throw your die once.

- Write the number in one of the nine single squares.
  Do this 8 more times.
  All the single squares should now have a number in them.

- Add up each row and put your answers in the double squares.

- Add up each column and put your answers in the double squares.

- What do you notice about square X?

3	5	1	9
2	6	4	12
2	3	1	6
7	14	6	X

- Now ask a friend to fill in some squares.

- The one who has the lowest answer in square X wins.

- Play again but make your own rules this time.

Teachers' Notes

21 Out

This activity should be preceded by 'hands on' experience of calculators, instruction on use, and discussion on functions.

Number

21 Out

Skills   Develop an ability to explore addition and subtraction patterns;
Solve problems using calculators;
Learn number facts to 20.

Materials   One calculator for each two players, pencil and paper.

Activities

- Write down the following numbers;

    1   2   3   4   5   6   7   8   9

- Each player takes it in turn to press one of the numbers and the key marked **+**.

- Each number must be crossed out after it has been used and not used again.

- The loser is the player who makes **21** or above.

Example

number of player	key presses	display
1	2 +	2
2	5 +	7
1	4 +	11
2	9 +	20

Player 1 will reach **21** or over with the next key press.

Player 2 is the winner.

Checking answers to computation problems gives practice in sensible calculator use.

Teachers' Notes

Take Six Dice

This activity is presented at both the representational and symbolic stages. It could be preceded by:

Physical experience

- Time could be set aside to 'play' with the dice, noticing the number of faces and arrangement of the spots. Other dice could be included e.g. with symbolic numbers and differing numbers of faces.

Communication

- This could include class discussion on findings from stage one.

Number

Take Six Dice

Skills Developing an ability to explain work and record findings;
Make and test predictions;
Know and use addition facts to 20.

Materials Six dice, pencil and paper.

Activities

- Take one die and place it on the table with the number one facing downwards.
  What is the number on the top?

- Take a second die and place it on top of the first one. The touching faces must add up to eight.

- If you continue to do this with all the dice, what will the number on the top be?

- Write down all the numbers you have used.

- Make some more towers and record the numbers you have used.

- Are there any numbers that you cannot use?

- Is there a pattern?

Teachers' Notes

Personal Data File

This activity is offered at the symbolic stage and could be preceded by a short task e.g.

Physical experience

Ask pupils to gather this information:

- Who is the tallest person in our class?
- Who is the shortest person in our class?
- What is the difference in their heights?

Communication

Questions to ask:

- How did you find the tallest person?
- How did you find the shortest person?
- How did you find the difference?
- Is there a better way?
- Did you measure in cm or m?
- How would you convert cm to m?

Representation

- Ask pupils to work in friendship groups to find the best way to record the information from stage one.
- As a class, examine the merits of all recordings and select one to be displayed on the wall.

Number, Shape, Space and Measures

Personal Data File

Skills Developing an ability to use a database;
Choose and use measuring instruments;
Use a calculator sensibly.

Materials One Personal Data chart per person, measuring tapes, calendar, calculator, bathroom scales, access to class number files (resource sheet page 74).

Activities

PERSONAL DATA FILE

My name is _____ Date of Birth _____

I am _____ years old
I am _____ months old
I am _____ weeks old
I am _____ m tall
I am _____ cm tall
I weigh _____ kg
I weigh _____ g

My Family

My _____ is _____ years old
My _____ is _____ years old
My _____ is _____ years old
My _____ is _____ years old
My _____ is _____ years old
Altogether we are _____ years old

My Class

There are _____ girls
There are _____ boys
There are _____ teachers
There are _____ assistants
There are _____ people in my class

My School

Class	Number of Pupils
_____	_____
_____	_____
_____	_____

In my school there are _____ pupils

Teachers' Notes

Down On The Farm

This collection of activities is at the symbolic stage. Many pupils will wish to use calculators.

Some pupils will need to use smaller numbers than 35 heads and 94 feet. Start with 2 heads and 6 feet, 4 heads and 12 feet and progress as far as possible.

Representation

- Pupils could be encouraged to use pictorial or concrete representation to help them solve the problems.

Number

Down on the Farm

Skill Developing an ability to solve problems using multiplication and division, use a calculator.

Materials Pencil, paper, calculator.

Activities

There are some hens and cows on a farm. Altogether they have 35 heads and 94 feet.

- How many hens?
- How many cows?
- If each hen lays one egg every day, how many eggs would the farmer get in one week?
- If each egg cost 10p, how much would the farmer get for his eggs in a day?
- How much would he get for his eggs in a week?

When the farmer puts his eggs into boxes each box holds **half a dozen eggs.**

- How many **full** boxes would the farmer get in one day?
- How much would a box of eggs cost?
- How many **full** boxes would the farmer get in **one** week?

Teachers' Notes

Take Six Dots

This activity is at the pictorial representation stage and could be preceded by:

Physical experience

- Present opportunity for pupils to use drinking straws or toothpicks to carry out this activity at the physical stage.

Communication

- Talk about the results of stage one in terms of horizontal/vertical/parallel lines.
- A walk around the school noticing lines brings meaning to the task.

Shape, Space and Measures

Take Six Dots

Skill Developing an ability to explore pattern.

Materials Cards, pencils, paper, scissors.

Activities

- Using three lines to join all six dots find how many different ways can you join them. Use a new set of dots each time. Each line must join two dots on opposite sides. Each dot may be used once only.

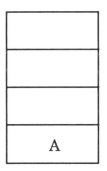
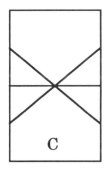
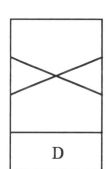

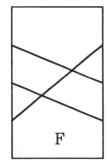

- What do you notice about: E and F?
  B and D?

- Cut out the rectangles and join them together by matching the lines.

Teachers' Notes

Telephone numbers

This investigation is at the symbolic stage.

Pupils who prefer to work with smaller numbers may wish to. Use the following approach to this investigation:

- Give each pupil a 9s line with the numbers 9, 18 and 27 already filled in; request that they fill in the next six numbers (using a calculator).

- Now ask them to make three cards each with a single number on them and write the numbers down in their work books.

- Next ask them to rearrange the numbers and record their new arrangement in their work books.

- Now ask them to subtract the smaller number from the larger number (using a calculator).

- Ask them to cross out any one number, except a zero, from their answer.

- Now ask them to add the remaining digits and make a note of the answer, then find the next highest number to it on their 9s line and find the difference between the two numbers.

- Ask pupils what they noticed about the answer.

- Ask them to try some more and see if there is a pattern.

Number

Telephone Numbers [1]

Skills Developing an ability to read, write and order numbers to at least 1000;
Explore number patterns and relationships using a calculator and make predictions about them.

Materials Calculator, pencil and paper, Telephone Directory.

Activities

- Take any three digit number, for example, 365. Then mix up the numbers (for example 536).
- Subtract the smaller number from the larger number.

$$\begin{array}{r} 536 \\ -365 \\ \hline 171 \end{array}$$

- Cross out any one of the numbers (if there is a zero do not cross it out), for example, ~~1~~71
- Add the remaining numbers

$$7 + 1 = 8$$

Now find the next highest number in the 9s line below.

| 9, 18, 27, 36, 45, 54, 63, 72, 81, ………… |

What do you notice?

- Try another number. Does it work again?
  Try some more numbers.
- Try some 2 digit numbers.
- Try some 4 digit numbers.
- Does it always work?

Number

Telephone Numbers [2]

Activities

Ask a friend for his or her telephone number, or ask him or her to find a number in the Telephone Directory.

Ask your friend to do the following:

- Write down the number so that you cannot see it.
- Mix the numbers up in any order and write down this number (to form a new number).
- Take the smaller number away from the larger number. Your friend may use a calculator.

Now, using the answer:

- Cross out any number except a zero from the answer.
- Add together the remaining numbers and tell you the answer.
- Now say, "I can tell you which number you crossed out."

Now look on your 9s line:

| 9, 18, 27, 36, 45, 54, 63, 72, 81, ... |

- Select the next highest number to the one your friend gave you.
- Take your friend's number away from that number and .... Hey, Presto ... there it is.

Example
• 553806
• 358065

• 553806
  358065
  195741

• 195741
• 1+9+5+4+1= 20 "I can tell which number you crossed out."
• Look at the number line.
• Select the next highest number - **27**
• 27-20=7 and that is the number crossed out.

Photocopiable Resources

PERSONAL DATA FILE

My name is _____ Date of Birth _____

I am _____ years old

I am _____ months old

I am _____ weeks old

I am _____ m tall

I am _____ cm tall

I weigh _____ kg

I weigh _____ g

My Family

My _____ is _____ years old

My _____ is _____ years old

My _____ is _____ years old

My _____ is _____ years old

My _____ is _____ years old

Altogether we are _____ years old

My Class

There are _____ girls

There are _____ boys

There are _____ teachers

There are _____ assistants

There are ____ people in my class

My School

Class	Number of Pupils
____	_____
____	_____
____	_____
____	_____
____	_____
____	_____
____	_____

In my school there are ____ pupils

Names and Numbers

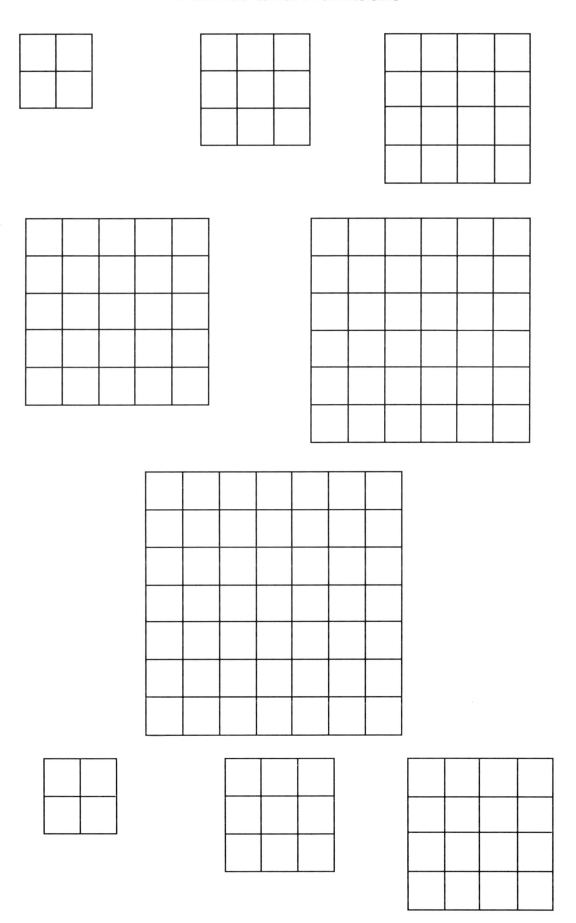

Grid Sums

Halving

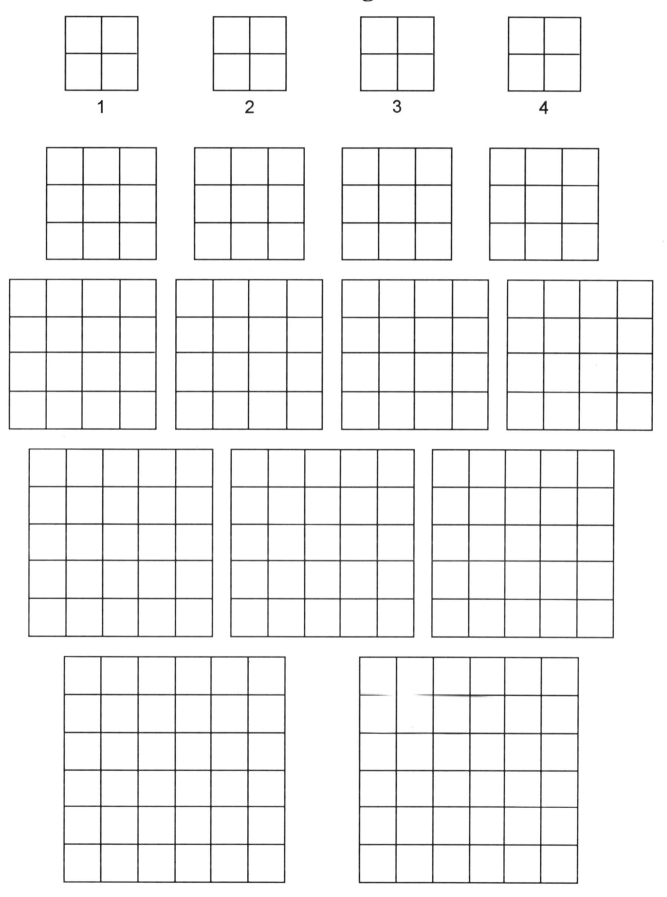

Houses

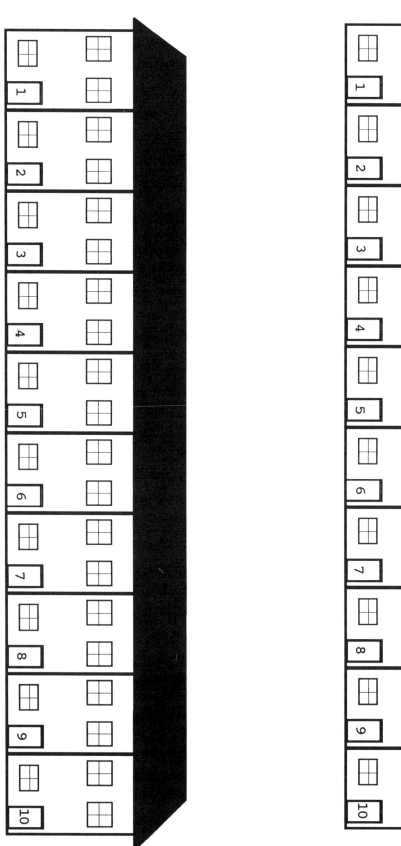
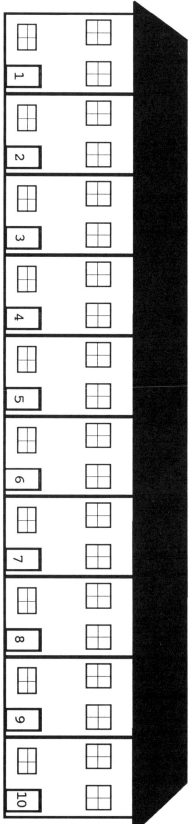

Making Patterns

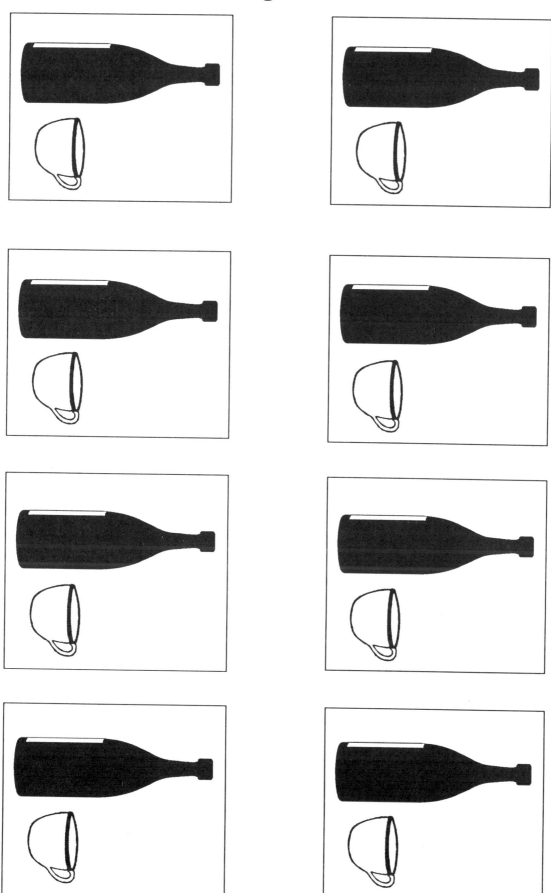

Shapes

Shape	Will it roll?	Will it slide?	Will it spin?	Build?	
sphere					
cube					
cone					
cylinder					
cuboid					